INHERITORS OF THE EARTH

The escape from water

Marshall Cavendish Books, London W1

Marshall Cavendish Books Limited
58 Old Compton Street, London W1

First published 1970
© Marshall Cavendish Limited 1968
© Marshall Cavendish Books Limited 1970

Printed by Proost, Turnhout, Belgium
462 00190 3

Picture credits

Contents

Introduction
by J.D. Carthy

Scientific Director, Field Studies Council

SO LONG as living things were tied to living in water, some twenty-five per cent of the world was barred to them. The ability to breathe air gave animals the possibility to escape their enemies by at least a temporary stay on land or to evade death by desiccation when the great droughts descended on the Earth. But new possibilities were opened when animals could live their lives on land — and reproduce there as well.

Armoured by their scales against loss of water from their bodies, fitted by their physiology to retain water within them and with eggs which could develop out of water, the reptiles could take over on land for, apart from insects, there were few other animals already there. They became large, they burrowed, some flew, others swam, but of all these great creatures relatively few descendants exist to-day. But as this book shows they represent even now the crucial changes which opened the land to the higher animals.

Some reptiles apparently developed feathers as modifications of their scales and the evolution of the birds was begun. Hugely successful as flying animals, birds have used their beautifully adapted wings to carry them into many places in the world — to the forests and the plains, back to the sea and rivers. Only an indication of the great diversity of bird life in the world to-day can be given here.

But the most successful set of animals to evolve from reptiles were the mammals. Taking on a constant body temperature and all the advantages that brings, they were able to oust the reptiles from their position of rulers of the ancient world. Their independance of the environment has permitted them to re-invade the sea and to live beneath the soil, as well as to climb in the trees and run on the ground. Some even fly. The adaptability of mammals described here is the cause of their success and no one of them more successful than man who can even use his technology to get him to the moon.

Reptiles — life on land

How they breathe and reproduce, their skin texture and limb development are features which mark the reptiles as one of the most critical stages in the evolution of the higher vertebrates.

LIFE WOULD HAVE no chance of reaching the level which we see around us in the higher vertebrates if it had not adapted in ways to make it possible to leave the water for dry land. The change was not easy, and Nature had two attempts at the job, both involving members of the animal group, the vertebrates.

From the first partially successful attempt, the amphibians, there stemmed the class of animals called *reptiles*. These have become fully adapted for life on dry land, although some still spend a great deal of time in the water. A hundred million years ago this class of animals was dominant, but outside pressures eventually brought the Age of Reptiles to an end. Now their numbers are reduced to a mere handful of types: the crocodiles, turtles and tortoises, snakes and lizards.

But before looking at the extinct or even the living species of reptile in any detail, we should first examine the general physical characteristics which have allowed reptiles their independence of water as a habitat.

From their amphibian forebears, the reptiles inherited lungs. These vary in sophistication, but in general they are superior to the lungs possessed by amphibians. In the case of the latter, the lungs are not efficient enough to supply the animal with all the oxygen it needs for life. Intake has to be supplemented by breathing through the skin. The reptile lung on the other hand has no such limitations.

Again, the reptilian system by which air is actually taken into the body is more advanced. The frog (amphibian), for example, gulps in air by expanding the muscles of its throat. These are then contracted and the air is pushed into the lung, the same air being used again and again until the oxygen content is exhausted.

The reptile, on the other hand, possesses a system of muscles which enables it to expand and contract its rib cage. This causes the lungs to work like a pair of bellows and breathing takes place in a more steady and controlled fashion.

This improved breathing apparatus has freed the reptilian skin from its breathing function and allows it to be used for another purpose, water conservation.

Since the skin of the amphibian is permeable to air, it also tends to be permeable to water and one reason why frogs and toads are clammy to touch is that they are permanently undergoing a process of evaporation of body liquids through the skin. Constant replenishment of this supply of water in the body is yet another reason why amphibians customarily live close to water. The reptile has overcome this problem by retaining a physiological attribute of a more distant ancestor, the fish. These are scales.

Reptilian scales are overlapping and consist of a hard, dry substance called *keratin*, which is virtually impermeable to water. It is therefore both a form of lightweight but effective protection to the body and also precludes loss of body fluids

The common iguana is a vegetarian and lives in South America. The throat sac, or *dewlap,* distends when the animal is excited.

by evaporation. The crocodile or snake, if provided with sufficient drinking water, is unlikely to suffer the fate of desiccation which so often befalls the frog and toad.

However in one other aspect reptiles are not well equipped – dealing with violent changes in temperature. It is obvious that beneath the surface of the water such fluctuations are much smaller than they are on dry land. This is concerned on the one hand with the physical nature of water and its potential for storing and losing heat, and on the other with changing seasons experienced on land where warm day inevitably gives way to relatively cold night temperatures.

Reptiles, like fish and amphibians, are what is generally termed 'cold blooded'. This is a splendidly inaccurate but widely employed term for what in science is called *poikilothermy*. What is really meant is that the blood heat of these animals tends to move up or down until it coincides with the temperature of their environment.

A very wide range of temperature would, however, cause irreparable harm to the delicate mechanism of the animal cells. To combat this possibility the reptile has been forced to develop ways of maintaining a fairly constant body heat.

In this the class has been only partially successful. Some lizards possess a reasonably elegant system by which they are able to change the intensity of their colouring in response to temperature change. They do so by contracting or enlarging the pigment contained in the cells in their skin. Thus they will become darker – with enlarged pigmentation – on cold days so that their skin absorbs as much of the available heat as possible, while on hot days the pigmentation is contracted and consequently the skin becomes lighter. The effect of this is to reflect the heat in much the same way as the white headcloth

The slow-worm, despite its snake-like appearance is a harmless, legless lizard. Here a female is surrounded by her young.

of a desert Arab.

This is only partially effective and most reptiles tend to hide from the sun when it is very hot and even burrow in the ground. In short they are better able to deal with cold than with excess heat, although they

Lizards are preyed on by snakes, weasels — and human beings. If part of the tail is lost, it will grow again, but with uneven scales different from those of the original tail.

7

Confronted with an adversary, the harmless grass snake puffs itself up menacingly; if this fails, it acts as though dead.

do need the average temperature of their surroundings to be fairly high in order to flourish.

Yet another important consequence for animal life of the change from a watery habitat to existence on land is that the body no longer has the support afforded by its buoyancy in water. For the animal to get about it needs limbs strong enough to support and move it.

Most reptiles, therefore, have four relatively powerful legs. (In the turtle these have become adapted for such purposes as swimming; in the snake, the legs have disappeared completely.) These limbs are paired, the forward set being connected to the bones of the pectoral girdle, the rear set to the pelvic girdle.

The legs are thought to be a development of the paired fins found in fish, and in bone structure they resemble the skeletal organization of the fins occurring in some extinct and some modern fish. In both, the fin or limb stems from one basal length of

The Indian python is also to be found in the Malay Peninsula. The female coils around her eggs and acts as an incubator.

Slender European whip snakes are extremely well camouflaged to live in the trees, their natural habitat.

bone; there are two bones in the 'forearm', several smaller bones in the 'wrist', and the system ends with a fan of five articulated bones.

But perhaps the greatest factor in the freeing of reptiles, and thus all the higher vertebrates, from dependence on water as a habitat is to be found in birth and reproduction.

As we know, the amphibian is forced to return to the water in order to produce its young. This is first and foremost because the infant needs water to support its life processes in both the egg and larval stages. Amphibian larvae are essentially fishlike; the tadpole, for example, has both fins for swimming and gills during its life up to the *metamorphosis* when it changes into the adult form.

Snakes and lizards use their tongues to touch and sense the surroundings. The tongue is flicked back into the mouth and inserted into a sensory pit behind the upper lip.

On the other hand, reptiles have evolved an egg which can be laid and hatched out on land. There is no fishlike larval stage; the young emerge from the egg as small replicas of their parents. In some cases the egg is held inside the mother until it hatches, and the young therefore come into the world live.

The importance of the development of the land-adapted egg is to make it even more unnecessary for reptiles to live where there are large stretches of water, and it is no surprise that they have been able to colonize, at one time or another, most of the land surface of the Earth.

But what are the requirements for an egg which will be successful on dry land? The first quality needed is without doubt a shell, which at one and the same time must prevent desiccation and also hold the contents together. The egg must con-tain sufficient food and water for the embryo to go through all its stages of development right up to adult shape before leaving the shell. Finally, it must have some means of storing or disposing of the waste products of the developmental process in such a way as to protect the embryo from poisoning.

All of these requirements are met in the reptile egg, one of the most elegant examples of natural design, a design which the reptiles have passed on to some mammals and birds.

It works like this. The egg is provided with a hard shell – unlike that of the fish and amphibia – and it is made of rigid calcareous material which at one and the same time is rigid enough to support the contents and is also impermeable to water. It is not, however, impermeable to gases.

Within the shell the infant is usually

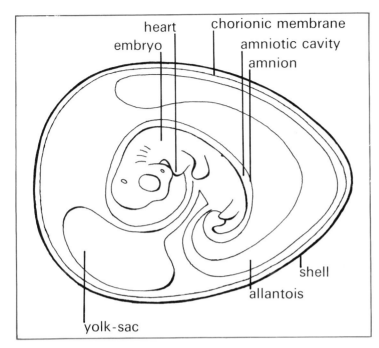

heart
embryo
chorionic membrane
amniotic cavity
amnion
shell
allantois
yolk-sac

Diagrammatic repres-
entation of a reptil-
ian egg shows the
four sacs or mem-
branes, the *chorion*,
amnion, yolk-sac and
allantois, which are
essential for the
growth of the embryo.

The two-month-old
embryo of a snapping
turtle is exposed by
removing part of its
shell. In later life this
variety spends most
of its time in water
and can be very agg-
ressive.

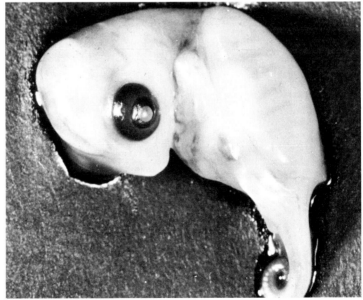

cushioned by a layer of *albumen*, of which almost 90 per cent is water. Next comes the embryo itself with its attached sacs, each of which has an important duty to perform. The *yolk-sac* provides the developing reptile's food; the *amnion* sac contains the embryo's supply of water in which it is constantly bathed. It is from this sac that this kind of egg gets its name. – the *amniotic* egg.

Caring for the embryo

The *allantoic* sac deals with all the waste products of the embryo. First, it holds liquid waste in its interior, and secondly, since it is connected to the circulatory system of the embryo, it is able to carry the gaseous waste to the inner surface of the shell through which it passes to the outside world. By the same means the sac collects in its blood system the oxygen coming in through the shell and passes it back to the

Hermann's tortoise is found almost exclusively on the northern coasts of the Mediterranean even though it is a land animal.

Turtles come ashore to lay their eggs. A 'nest' in the ground is excavated and the eggs placed in it. The mother then leaves the job of incubation to the warmth of the earth.

embryo. In this sense it is both a respiratory organ and a waste-disposal unit.

Finally, the *chorion* is the sac which surrounds both the *amnion* and the *allantois*. It assists the former in the job of providing the aquatic medium which is so necessary to the development of the embryonic reptile. Further, though the allantois and shell are separated by it, the chorion still allows the gaseous exchange to take place.

The amniotic egg is a complicated arrangement, but although it frees the reptile from the need to breed in water, it has required one other adaptive change in these animals. The hard shell of the eggs makes it necessary for them to be fertilized before that shell is formed – while the egg is still within the mother. Most male reptiles, therefore, possess some type of

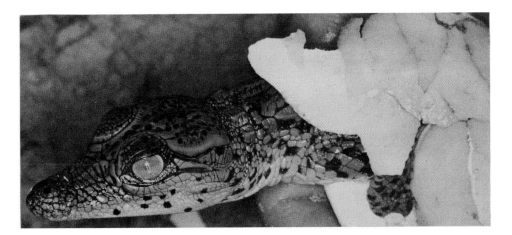

Crocodilians lay clutches of about 30 eggs. At the end of its snout the emerging crocodile has an *'egg-tooth'* for breaking out of its hard, waterproof shell.

intromittent organ, or penis, for the injection of the sperm.

The modifications so far discussed are backed by changes in the nervous and sensory systems. Such changes are necessary for the simple reason that without them the reptiles would be unable to make full use of the other faculties which make fully terrestrial life a possibility.

Crucial among these developments is the sense of hearing. In simple terms the reptilian ear, like that of other land vertebrates is made up of two parts: the sense organ itself and the means of transmission of the sound waves.

The device for performing the latter function is called the *tympanic membrane,* or *eardrum.* In many reptiles this membrane is visible on the side of the head. In others it lies protected in a pit.

What happens is this; the airborne sound waves beat upon the taut tympanum and are then transmitted across a cavity called the *middle ear* by a slender bone, the *ossicle.* Travelling along this the signals finally reach the second part of the hearing mechanism, called the inner ear. This is where the actual hearing is done in the sense that the sound waves are turned into signals which impart information to the brain.

Again, the eye of the reptile is adapted to do the job of seeing out of water where the refractive index is different and for that reason exhibits significant differences from the eye of the fish. For example, the *cornea* in the fish eye is nothing more than a transparent window. It has no function in focusing the image on the retina. In the reptile the cornea has become curved and plays a considerable part in image formation.

As the reptile eye is now out of water for a good deal of the time its surface must be kept clean by some mechanism in the animal itself. The reptilian eye has therefore developed eyelids – usually three in all. The first two are an upper and a lower lid to each eye, of which the lower is the larger and more mobile. However, it is the third lid which is thought to do the cleaning. This is almost transparent and sweeps backwards from the inner corner of the eye across the surface of the cornea. In mammals this third eyelid has degenerated and exists only vestigially in the inner corner of each eye.

As a footnote, the colour of the reptilian eye is often bright red or yellow, and it has been shown that many of them have the ability to distinguish between colours, although, for some reason, this is thought not to apply to the crocodile.

One per cent brain

The move from water has also brought about the disappearance of the nervous systems associated with gill breathing and also that peculiar organ found in fish called the *lateral line system*. This is generally accepted to be a means by which fish receive information about changes of pressure in the water around them, and by which they may even locate objects in their path and maintain formation with other fish in their particular school. Clearly the terrestrial reptile has no further use for such mechanisms and consequently they have disappeared.

Although the brain shows some development in terms of size from that of amphibia and fish, it remains relatively small in comparison to that of mammals and in comparison to the reptile's own size. It is unlikely to amount to more than one per cent of the animal's body weight. Even in the mighty dinosaurs, some of which weighed as much as 20 tons, it is unlikely that the brain accounted for more than a few ounces and was probably only a few inches long.

All these adaptations have enabled the reptile to survive without dependence on an aquatic environment. Without ignoring the variations and peculiarities which characterize individual types within the reptile family, the physical attributes of this class of animals sets them apart from their antecedents to form perhaps the most important link in the evolutionary chain which has led to the higher vertebrate life on Earth today.

Reptiles –
the defeated conquerors

The kings of the Earth, the mighty dinosaurs, have long since faded into extinction. Only four of the original 16 orders have survived, but reptiles still inhabit most corners of the world.

SAY 'PREHISTORIC ANIMALS' to the average person and they immediately think of the dinosaurs. Yet even then they are not usually thinking of dinosaurs in general but of the giant *Brontosaurus*, merely one of the group. There is little notion of the range and diversity of the reptiles which existed, or that the way the number of species have dwindled is one of the most dramatic happenings in natural history.

The Age of Reptiles – that is the period during which they dominated the land areas of the Earth – was in fact the whole of the Mesozoic era, which began about 205 million years ago, although reptiles did exist in some numbers before that. In fact, they began their climb from origins among the *labyrinthodont amphibians* much earlier in the Paleozoic era.

The dinosaurs were without doubt the most impressive of the reptiles in the Mesozoic era, but they were not only large like the brontosaurus.

The savage *Tyrannosaurus*, about 50 feet long, walked on its hind legs and was

a meat eater. Its enormous and powerful head was equipped with sharp sabre-like teeth, but its tiny front legs were rather feeble in marked contrast to the rear ones. The *Stegosaurus* (the name refers to the armour plates which ran down its back and suggest roofing tiles) had only weak teeth and jaws and is considered therefore to have been a vegetarian. Also in contrast to the tyrannosaurus it walked on all fours. Some reptiles moved towards exploiting the air; these were the *Ptero-saurs* or winged reptiles, which though they did not lead directly to birds as we know them, were certainly successful for a time.

That the pterosaurs could fly was due to the extraordinary development of their forelimbs. In the pterosaur the fourth finger of the forelimb became enormously extended, the fifth finger disappeared and an area of skin stretched from this extension to the sides of the body to form the wing. Some were no bigger than modern bats, although others boasted wing spans of up to 29 feet and were the biggest animals ever to fly.

Birds as we know them also emerged from a reptilian root, the *thecodonts,* which also produced the dinosaurs. In its early life this line progressed less quickly than that which produced the pterosaurs but ultimately it emerged superior.

But what brought about the end of the Age of Reptiles? Why were the number of reptilian species so reduced at the end of the Mesozoic era? At this distance of time it is of course extremely difficult to answer these questions, and many of the reasons given can be dismissed as fanciful. There was probably more than one causal factor.

First, it may be that the temperature of the Earth rose towards the close of the era and all reptiles found difficulty in dealing with increases in temperature. There is fossil evidence that plant life took on a sudden spurt in the later Mesozoic and this could be accounted for in climatic terms (which would support the temperature-increase theory), but it might also be due to geological changes.

However, if the climate did fluctuate in this way, it may be that the reptiles perished simply because things got too hot for them. With the evaporation of the lakes the animals would be without facilities for hiding from the sun.

The problem of heat

It is likely that the larger reptiles disappeared leaving the relatively smaller forms of life, because the greater the surface area which the animal presents to the sun, the bigger are its problems in terms of resisting overheating. Many could not overcome this inability to deal effectively with temperature fluctuations outside the water.

Modern *Reptilia* are represented by four different orders. The three most common are the *Squamata* (lizards and snakes); the *Crocodilia* (crocodiles, caimans and alligators); the *Chelonia* (turtles and tortoises). The fourth order is the *Rhynchocephalia* and this has only one surviving species, the Tuatara which is found only on small islands of the North Island of New Zealand. Of the four orders, the Squamata is by far the largest. There are over 2,000 species of lizard and about 2,400 species of snake.

The typical lizard has an elongated body with two pairs of limbs terminating in five-toed feet. Some less typical kinds, however, resemble snakes more closely in that they have no limbs at all, or at best merely rudimentary ones.

The difference between lizards and snakes lies firstly in the distribution of scales. The snake has transverse scales across its belly, while in the case of the lizard that area is covered like the rest of the body in small horny scales. Secondly there is the formation of the jaw. A snake is able to swallow meals much bigger in diameter than itself. This is because the jaws are cunningly articulated, the lower one being split into halves connected by rubbery tissue. The lizard has no such ability.

In one way, however, all lizards are snake-like. This lies in the way their bodies curve as they move across the ground, and is a function of the way the legs are constructed. Where legs do not exist, the lizard gets about in even more of a snake-like style.

Most of the sub-order live by hunting and are flesh eaters. Those that do eat vegetable matter are not strictly vegetarian.

The lizards have an interesting trick or two up their sleeves when it comes to survival. Some can go without water for extremely long periods, slaking their thirst by sucking the dew from stones. Some varieties are able to lose their tail when seized by another animal leaving the lizard to scamper off to safety, apparently unharmed. The reason for this is that such animals have a disc of gristle

A mouse is swallowed whole by a boa. The flexible articulation of the jaws enables the snake to engorge prey of much greater diameter than itself.

Largest of the monitors is the Komodo Dragon, which can reach ten feet in length. It has an enormous appetite and eats small deer, wild pigs, carrion — and is also cannibalistic.

inserted into the middle of their tails, between two of the vertebrae. The muscles and veins in each part are so designed that they will easily break apart at this point and the damage is minimal. Later a new tail will grow, but it is unusual for it to reach the length of the original. Because of their survival equipment, lizards are to be found all over the world, except in polar regions. However, they flourish most happily along the equator and de-crease in numbers progressively towards north or south.

The sub-order of lizard is itself divided into several families, and of these the *monitors* contain the largest types. The greatest of all is the Komodo Dragon which can weigh up to 290 pounds and can be over ten feet in length. It was first dis-covered in 1910, and lives on Komodo Island, which is east of Java and a mere 18 miles long by 12 wide, and a few smaller

1 The Gila Monster is one of only two known poisonous lizards. It lives on small birds, eggs and young mammals, but can fast for long periods after storing food in its tail.

2 Tiny hooks on the undersides of the gecko's toes enable the animal to run very fast and cling to almost any surface. The name is derived from the repetitive 'geck-o' call of some species.

3 *Chamaeleo bitaeniatus* inhabits the mountains of eastern Africa. Chameleons possess a prehensile tail for climbing, a prehensile tongue for catching insects, and has all-round vision.

islands nearby. In spite of their large size, they are capable of a good turn of speed. Usually they remain on land, but can swim, and they burrow a hole in the ground in which to rest at night.

The *Helodermatidae,* another family of lizards, boasts some formidable specimens of its own. These include the only two existing species of poisonous lizard. One at least, the Gila Monster, has a bite which can be fatal to Man and very soon dispatches small mammals. It is a sluggish animal which moves mostly by night, and grows to nearly two feet in length. When disturbed it hisses loudly and froths at the mouth.

Chameleons and geckoes make a pleasant change from the rather unattractive monsters already mentioned. In the former the body carries a sharp ridge down the back and is best known for its ability to change colour to harmonize with its surroundings. In a rage it turns deadly white. The gecko is very common around the Mediterranean and is most notable for its ability to run across the ceiling of a house with its 'hook-pad' feet.

The snake, the second sub-order of Squamata, is a beast for which Man seems to feel an innate repulsion, but only about 250 species out of a total of some 2,400 are in fact poisonous to Man. All snakes have the jaw mechanism already mentioned which allows them to take in large quantities of food at one time. This in turn means that they eat at fairly widely spaced intervals.

The teeth of the snake are thin, pointed and tip backwards towards the throat. In the poisonous varieties the venom sac is situated in the upper jaw, and the associated teeth have either grooves running down the outside or a canal running through them to carry the poison to the bitten victim.

These creatures, of course, have no legs; however, they do have a great many ribs – commonly a pair to each of as many

A blue-throated tree agama. The *agamids,* similar to the iguanas, comprise 200 species of lizard found in warm parts of Asia, Africa, Australia and southern Europe.

of 200 vertebrae. They use the tips of these ribs to get them about much as a millipede uses its vast number of limbs. Snakes slough (discard) their skins at regular intervals throughout their lives.

The boas and the pythons are non-poisonous snakes but are well known for the manner in which they kill their prey. Both grow to a considerable length and attack by striking at the victim with their jaws and then squeeze the life out of it. The largest boas are probably the South American anacondas, which have been recorded as growing to as much as 37 feet in length.

The major difference between the boas and pythons is that the former bears its young alive, while nearly all the latter lay eggs. The python, however, provides perhaps the most spectacular example of the

The Egyptian cobra, although smaller and less dangerous than his Indian brothers, is still a man-killer. These snakes kill by ejecting venom into their wounded prey.

snake's ability to eat animals much bigger than itself in girth. One nine-foot Reticulate Python (from South East Asia) was observed to swallow a monitor five feet long; another dealt with a deer weighing 123 pounds.

Eating under water

The crocodilians (crocodiles, alligators, caimans and gharials) are examples of reptiles which live most of their lives in water. However, they are fully equipped to move about on the land and commonly do so to find new stretches of water and to lay their eggs.

All species closely resemble one another. They have long heads with powerful jaws and tails for swimming. Their legs carry five toes on the front pair and four on the back and the toes are usually webbed.

The crocodilians are able to lie submerged and yet open their mouths to take in their prey. This is due to a physiological mechanism consisting of a fold in the tongue and a flap hanging from the roof of the mouth. When these are brought together they prevent the lungs and bodies from flooding even though the mouth is open. The largest of the group are probably the gharials (or gavials) of which there is only one species, found in the Ganges. It is characterized by a very long and slender nose which contains teeth finer than those of other crocodilians. Gharials have been known to grow to 21 feet but they do not attack Man.

The true crocodile has a shorter snout than the gharial and exists in most continents other than Europe. It is prepared to attack Man and even larger animals, should they venture too close.

The alligators and caimans comprise a

Although they have the longest snout of all crocodilians and grow up to 20 feet long, gharials are not dangerous to Man. They keep almost exclusively to their diet of fish.

Crocodiles lie submerged with only their eyes above the surface. They can eat in this position – folds of skin inside the mouth prevent them from gulping large quantities of water.

The leathery turtle is the largest marine turtle. A fully grown adult may be six feet long and weigh up to 1,000 pounds. A female returns to the water after laying her eggs.

family of seven species of which all but one live in America. The odd one out is found in the lower reaches of the Yangtse-Kiang River, China, and has been studied relatively little. One peculiarity is that its toes on the forelimbs are not webbed. This may be due to the fact that it spends the winter in a hole which it digs in the mud – presumably any webbing on its front feet would become torn in that process.

The black caiman which is abundant

along the Amazon departs from the usual method by which crocodilians get their meals. Instead of seizing the victim in its jaws and dragging it to the bottom, the black caiman stuns the fish or other animal with a swipe of its powerful tail. They have been known to leave the water during the night for hunting trips, killing sheep and small animals.

The Chelonia (tortoises and turtles) are particularly easy to distinguish from other forms of reptiles since they carry a *cara-*

A resting place for a large number of 'grinning' American alligators. The short, broad head dis-

tinguishes the family *Alligatoridae* from the slimmer and longer-headed crocodiles.

The European pond tortoise is a strict carnivore feeding on fish, frogs and insects. In autumn it hibernates by burrowing into the mud and remains there until the spring.

Almost a 'living fossil', the Tuatara is the only survivor of a group of reptiles common during the Mesozoic era. Today it lives on a few islands near New Zealand and is protected.

pace or shell which is usually hard but not necessarily so.

They all possess four limbs and breathe in the characteristic reptilian manner through systems of muscles forcing their lungs to contract. The aquatic turtles can take oxygen from the water and emit carbon dioxide through the skin.

Chelonians can live to 100 years old. They live in the tropics, although a few are found in temperate regions. They usually spend the winter in hibernation.

Finally, the Tuatara, which resemble lizards to the casual observer, exhibit internally a number of features which show them to be related to more extinct forms. It is mainly nocturnal in habit and moves about slowly feeding on worms, insects and frogs.

In all of this we can see the decline of a once dominant class of animals and some of the reasons for its decline. However, the modern forms which remain are well adapted to the lives which they lead, and it is likely that without some disastrous intervention by Man they will survive.

Birds — feathers and flight

The combination of wings, beaks and feathers sets the birds apart from other animal classes. From reptilian beginnings each has adapted to its own environment — even reverting to flightlessness.

LIFE BEGAN in the oceans; after millions of years of development it progressed to the land and diversified slowly. At this point there only remained one area suitable for the support of animal life, the atmosphere. But progress from Nature's drawing-board to a living flying machine was slower and more complex than any process imaginable in terms of aeronautical engineering.

The birds, a branch of the vertebrates, have their origins among a group of reptiles called the *thecodonts* which were able to run about on their back limbs; the front limbs of one group of animals developed into wings.

Learning to fly

First to appear were the *pterosaurs* which were truly winged reptiles. Their wings were not as sound in aerodynamic terms as those of the true birds and were much more easily broken.

The structure of their legs from fossil evidence seems to suggest that they had difficulty in taking off from flat surfaces, and would in any circumstances have been ungainly on the ground. Though the largest animals ever to fly must be counted among the pterosaurs, they represented something of a blind alley in the progress towards unhindered flight, and died out with most of the other reptiles at the end of the Mesozoic era.

Perhaps the most important acquisition of the animals comprising the second line of development was the feather. It is widely accepted that feathers themselves evolved from reptilian scales, but the change could hardly have been more drastic. The only similarities are that they, like the reptilian scales, are made of *keratin* (a substance produced by the skin, also responsible for hair and horn), and that they both occur in a regularized pattern over the bird's body and wings. The modern bird, incidentally, has retained scales on its legs.

Unlike scales, feathers are enormously intricate in construction and it is worth looking at them in some detail. Each feather has a quill or main shaft from which extend some 600 hair-like barbs on either side. But the resemblance to hair is extremely superficial, for each of the 1,200 offshoots carries about 800 smaller ones,

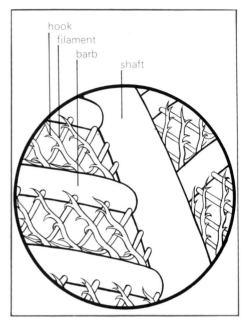

The microscope reveals that a feather is not merely a shaft with barbs extending from it. Filaments with minute hooks lock together to form an almost airtight mesh.

and each of these smaller offshoots carries as many as three dozen tiny hooks. The whole of this complex is locked together by means of these hooks, and the marvellous interweave which results is so finely meshed that air can only just pass through. If you pick a feather from the ground and handle it roughly all the smoothing in the world will not return it to its original shape. To fit all those tiny hooks accurately together again is just not possible.

Of course in normal life the feathers of the bird do become spoiled. But they, in some unknown way, are able to re-engage the hooks by rubbing the feathers with their beaks. Eventually the feathers become irreparable and drop out during a period which occurs once or twice a year in most birds and is called the *moult*. Since the feather is so important to flight, birds tend to become less active during the moult, which lasts until the lost feathers are replaced by new ones.

Associated with the feathers are muscles which enable the bird to raise and lower them. They are, therefore, important for flight, but more relevantly at this point they also play a vital role in regulation of the bird's body heat.

Living cells are extremely vulnerable to damage by fluctuations in external temperature, so the body temperature must be kept fairly even. In the reptiles, including the extinct pterosaurs, the mechanism of heat control is not very effective; they rely on moving into shade when it is hot and the open when it is cool. Obviously then, outside factors can wreak havoc upon a reptile population; drought can wipe out suitable shade with the sudden destruction of vegetation.

By means of its feathers the bird is able to overcome this heat control problem. The feathers trap air beneath them and, as with the large-mesh string vests used by Arctic explorers, this layer of air becomes heated and proves an effective method of insulation from the cold.

But even so the usefulness of the feather does not end here. They are vital to successful flight since the way they are distributed over the body and wings produces effective streamlining. The main supporting surfaces of the wings are comprised of long flight feathers. Banks and turns impossible for an aeroplane are open to the bird through movements of particular sections of the wing feathers.

Wind and weather proof

Feather colouring is important both for purposes of camouflage and mating. Feathers are also associated with both waterproofing and the sense of touch. Most birds have two oil-producing glands situated near the tail; the bird when preening presses its beak against the glands, whereby it receives a coating of oil which it then transfers to the feathers, so enhancing their ability to repel water.

1 In America the swallow is known as the barn swallow because of the bird's close association with Man and his environment. It often builds its nest in man-made nooks and crannies. **2** The strong, sharply curved beak of the peregrine falcon is well suited to the job of killing and dismenbering its prey.

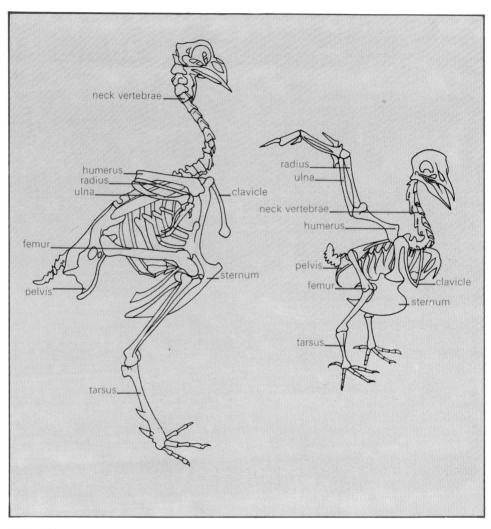

Flying birds have a relatively light skeletal frame with strong and well-developed wings (right). The hen (left) is an almost flightless bird with small wings and a bulky body.

What is not clear is how other species – some parrots, for example – manage without oil glands at all. As to touch, nerve fibres around the base of each feather provide birds with extensive sensitive areas.

Even though the feathers are crucial to both flight and temperature control, they would have been of little use in the development of the modern bird without other extensive physiological changes from the reptilian pattern.

The wings, as in the pterosaurs, depend upon the adaptation of the bone structure of the reptilian forelimbs. The wing is supported first by the upper arm (the humerus), then by the forearm (the ulna). The 'wrists' are relatively simple and the skeletal system terminates in a much extended second finger. The supporting bones are hollow and reduced to a minimum while still retaining sufficient rigidity for action; lightness is absolutely vital. An indication of the kind of weight saving which is achieved is that the frigate or man o'war bird, although having a wingspan of more than seven feet, has a skeleton which weighs a mere four ounces. Apart from the wings, the rest of the skeleton consists of light, hollow bones and is suitably adapted to the bird's natural element. The breastbone, or *sternum*, extends along the bottom of the animal's body – from the base of the neck to the tail – like a thick knife-blade. Strengthened also by ribs and backbone, it forms a massive and rigid anchor for the muscle system of the wings.

When not in the air, the bird is able to walk or hop about the ground. It can do so because of the formation and positioning of its leg bones. The legs are toward the front of the body which means the bird's centre of gravity is situated over them; the thigh bones slant forwards and, in conjunction with the muscles of the legs, also help balancing. Thus a bird such as the flamingo is capable of standing easily on one leg, and even of going to sleep in that state.

Feet consist in most cases of four toes, three pointing forward and one back. This configuration makes it easier to move fast on foot and explains the advantage which the modern bird has over the extinct flying reptiles on the ground.

Apart from containing a relatively large brain, the bird's head is dominated by the beak, or bill, which is essentially a tool for seizing and handling food. The neck

1 A coot's nest is built among the reeds, often floating on water. The six to nine eggs are brooded by each of the parents in turn. 2 Puffins, once on the verge of extinction in Britain, are now protected birds. They are equally at home on land or in water, but are particularly good at swimming and diving for food.

1 The most common owl in Britain, the tawny owl, has plumage consisting of fine feathers of varying colours which confuse its enemies. **2** The family *Paridae,* of which the long-tailed tit is a member, build intricate spherical nests of moss and lichen. Through the round entrance a female feeds her young. **3** Long, pendulous nests hanging in groups from a tree belong to weaver-birds. On closer inspection the intricately woven patterns become more apparent.

normally carries many more vertebrae than are found in mammals, and consequently has considerable mobility. In fact, such power of movement is essential in conjunction with the bill for the gathering of food.

The bird's muscular system needs to be supported by an equally powerful and efficient system of circulation and respiration. The heart is divided into left and right sections, and relatively speaking is much larger than the heart of the mammals.

Blood, breathing and brain power

The right section pumps blood to the lungs where it takes up oxygen; the left side pumps the oxygenated blood at high pressure to the muscles of the body. The red blood cells – those which actually carry the oxygen – are capable of transporting large quantities of the gas which they are able to give up very quickly, thus keeping high the oxygen level of the tissues.

The lungs of the bird are relatively small, but that does not mean to say they are not highly efficient. The reason is that they are supported by thin-walled air sacs which help once again to lighten the body and even extend into the actual bones of the animal.

The intake of air is greatly increased in this way beyond the capability of the lungs themselves. On breathing out – expiration – it is thought that the air in the sacs passes through the lungs again where further gas exchange takes place within the circulatory system. It also makes sure that stale air is cleared fully from the lungs themselves.

The brain of the bird is well developed; it has very effective senses of hearing and eyesight. Birds react to other bird calls over fairly large distances. Likewise the excellent eyesight of this class of animals can be gauged from the way in which they are able to spot airborne predators, hawks and buzzards, when they are flying very high or a long way off.

On the other hand the sense of smell in birds is not so good. The sense of taste is good in some varieties; most will reject strongly salted food.

A bird's song is one of the greatest

pleasures which animal life can give us. The voice-box of the bird is a complicated system of membranes and cartilaginous rings, and these are activated by the column of air drawn down the windpipe.

The sounds which a bird is able to emit are of the utmost importance to it. They enable the adult animals to warn of approaching enemies, to attract mates, and to let intruders know that they are entering a particular bird's territory; the young can keep in touch with their parents and show when they are hungry. The birds have inherited from their reptilian ancestors the kind of egg necessary for successful reproduction on dry land. First and foremost among the requirements is that such an egg should have a hard shell which is capable of holding the contents together. A natural consequence is that the egg must be fertilized by the male before it leaves the body of the

1

2

3

4

5

1 An outsized baby for this unfortunate hedge-sparrow. The cuckoo lays eggs in the nests of other birds; when the young bird is hatched it pushes out the rightful occupants and continues to be fed even when it grows larger than its 'foster-parents'.

2 Green woodpeckers cling to the tree with their claws and inspect the bark for insects with their long beak and tongue. Nests are built in a hollow tree-trunk.

3 A brooding red-legged partridge is disturbed in the long grass. These birds lay 10 to 16 eggs in a depression in the ground.

4 Found near lakes, the black and white Australian mudlark constructs its nest with mud reinforced with hair, feathers and grass.

5 Oyster-catchers in profusion. Their long red bills are well designed for opening mussels, limpets and, occasionally, oysters. The female lays eggs in the smallest of hollows.

female, and in some cases the male bird possesses a penis although more usually the sperm is transferred by external bodily contact. Hatching is sometimes aided by the parents sitting on the eggs; sometimes, as in the case of the cuckoo, by other parents. In warm climates, however, the mother may simply leave her eggs behind and allow the sun and the warm earth to do the job.

Periods of incubation vary enormously. In some smaller birds it can be less than a fortnight; in the larger ones such as eagles and albatrosses it can range from two to three months. The variety of nests built for the eggs is enormous. Some merely use the ground, others build the classic cup shape of mud and interwoven grasses, and yet others rest their brood on nothing more substantial than a pile of twigs.

Clearly the animal which can fly has a number of advantages over his land- or water-locked compatriot. When danger threatens on the ground it can move at high speed to safety. If the right kind of food or vegetation is destroyed in an area, the bird can take to the wing and look for new pastures.

It is, therefore, something of a surprise that some species of bird are flightless, while retaining all or most of the other characteristics of the breed. It might be thought that these were types which had not progressed so far along the evolutionary road, but this is not so. In fact, they have degenerated from types which *could* fly but have since lost the ability.

White storks, once paired, remain together for life; they build their nests, frequently in exposed places, and return to them each year, gradually adding more and more material.

Why this should have happened is not known for sure, it is only possible to guess. For one thing it has been found that many flightless birds occur or have occurred on islands, and the likelihood is that they were not bothered by mammal predators. In New Zealand, where there are no native animals of this type, not only the kiwi but also the duck, owl and penguin are flightless.

Flightless and flying high

For the ostrich, a native of Africa, this theory does not hold true because the continent has perhaps more than its share of predatory mammals. In this case it has been suggested that the large size of the birds, and perhaps their fleetness of foot, has allowed them to live on without the ability to fly.

Whatever the reason for the flightlessness of these animals, they are mavericks or degenerates from the normal. Birds have achieved their extraordinarily wide distribution in the world because they are able to fly. They have conquered the skies in a way that Man, even in a modern aeroplane, can never hope to emulate.

Birds —
soaring far and wide

Almost 9,000 species of bird live in the world today. Through their ability to fly, birds have inhabited almost every corner of the world — even to the edges of the great polar icecaps.

'FREE AS A BIRD.' How many times must these words have been used to describe a blithe and carefree state, shorn of all depressing restrictions? In one sense, the expression represents our wistful yearning to be able to get away from it all with as little trouble as these flying vertebrates apparently experience. It also reflects, by implication, the admiration which Man has felt from time unremembered, for the grace and efficiency with which the birds exploit the skies.

Nobody can forget the majesty and power in the wingbeat of a circling Golden eagle or a transitory swan. Smaller birds – in particular the tiny, colourful hummingbird possess a manoeuvrability which is breathtaking.

The internal structure of the bird is designed for lightness. It has slender and hollow bones, and its lungs are supplemented by air sacs. Similarly its feathers, while important for maintaining body heat, are also light and effectively streamline the bird's body. But it is the wings which really do the job of raising the bird into the air and maintaining it there. They provide lift and propulsion unlike the wings of the aeroplane which provide only lift.

The important wing movement to gain altitude is the downstroke, for it is during this movement that the under surface of the wing presses down upon the air beneath it. Naturally the air offers resistance to the wide-spread wing, and this levers the bird upwards in much the same way that oars propel a rowing boat.

The most important part of the wing in this context is the tip. Its vertical arc is greater than that of the rest of the wing and, like the blade of the oar, it produces the greatest amount of leverage.

It is important for propulsion. For forward flight, the leading edge of the wingtip is held slightly lower than the trailing one, which means that the tips not only press downwards against the resistance of the air, but backwards as well. The analogy with oars is even more fitting in this case.

When it comes to the upstroke of the wingbeat cycle, the front edge of the wing is raised relative to the trailing edge. As the bird is still moving forward by the force of the previous downstroke, the air through which it is passing builds up against the underside of the wing. At the same time the wing assumes a delicate curve so that the flow of air across it is smooth and as little speed as possible is lost. This situation cannot last long. In most conditions the bird will be gradually descending and will continue to do so, unless it begins the wingbeat cycle again, with another downstroke. What is called level flight, is nothing of the sort when it comes to bird movement. In fact, the creatures progress in a series of arcs, though these may be so short that the flight path may appear level.

Soaring and gliding

There are important areas of flying which do not depend upon the flapping of the wings, but on spread enabling birds to soar and glide. When gliding the wing is held in the same configuration – that is, with the leading edge held higher than the trailing one – as in the upstroke of level flight mentioned above. But, since in this case there are no downstrokes to maintain

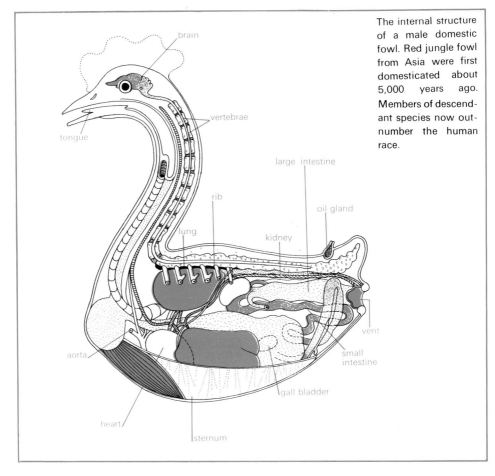

The internal structure of a male domestic fowl. Red jungle fowl from Asia were first domesticated about 5,000 years ago. Members of descendant species now outnumber the human race.

brain

vertebrae

tongue

large intestine

rib

oil gland

lung

kidney

aorta

vent

small intestine

heart

gall bladder

sternum

The northern gannet lives around the northern shores of the Atlantic. It hunts fish and is an excellent diver and flyer. Although nesting in colonies each pair has only one egg.

height for the bird, it must sink, however gradually, even though it moves forward.

Soaring involves the use of the wings solely as sails. They perform a relatively passive role and are used merely to collect the pressure of air currents which are moving upwards from the surface of the Earth. The upcurrents consist, in one form, of air which is heated at ground level by various means – a town, for example, emits a good deal of heat – and consequently rises. These are called *thermals*. Upwelling occurs when a surface breeze rushing across the ground is deflected upwards by meeting a cliff or other obstruction. Either way, the birds are capable of employing these columns of air to carry them to great heights.

The formation of the wings of a bird clearly reflect the type of flying it does. Creatures like the buzzard and frigatebird have long and broad wings because they spend a great deal of time gliding or soaring. A long wing cannot be moved through its arc as quickly as a short wing with the same expenditure of power, so birds which depend for a living on very quick and agile movement – the sparrow hawk, for example – have wings which are short and broad. Because air will not flow smoothly over a very broad wing at speed, the fast flying birds tend to have narrow swept-back wings.

The highest flyers seem to be found in the region of the Himalayas; the alpine chough, a rather dull-looking black bird, has risen as high as Everest – about 29,000 feet. Other champions from the same area are the lammergeier, which has reached 25,000 feet, and the red billed chough and wall creeper which have attained 21,000 feet.

Fantastic claims have been made for the speeds which birds have achieved in the air. As much as 200 mph has been attributed to the brown-throated spine-tailed swift of Asia, but the observer was not equipped with reliable timing apparatus. In all such cases the claims emerging must be treated with caution. However, it has been proved conclusively that peregrine falcons have achieved as much as 180 mph when swooping on their prey, but here it might be said that they are falling rather than actually flying.

Humming-bird 'helicopter'

In level flight the racing-pigeon has reached a creditable speed of 94 mph. Other speedy birds are the loon (90 mph) and once again the high flying lammergeier (79·5 mph). Most birds, of course, never exceed 40 mph, but an interesting exception is the ruby-throated humming-bird, which can hover like a helicopter – maintaining its position with 55 wingbeats a second – and fly at 60 mph.

Some birds habitually cover enormous distances on their migratory flights, the journeys which are, it is thought, prompted

To discover migration routes and other bird movements, ornithologists ring the legs of selected specimens - in this case it is the common black-backed gull.

The male whinchat calls for a mate after returning to England from winter migration. A female that likes the bush the male has chosen will nest with him beneath it.

by changing climatic conditions. 'One swallow does not make a summer', but when there are plenty of them around it usually means that the British summer is well under way. Few people realize, however, that these relatively tiny birds have flown all the way from Africa. The Manx shearwater has been known to travel 3,000 miles and the white-fronted goose commutes from Greenland to Ireland, a distance of some 2,000 miles across the choppy featureless surface of the Atlantic.

There is some evidence to suggest that birds navigate by the sun during the day and the stars at night, but exactly how the brain uses this positional information as a basis for action is not at all clear. At other times it is known that the creatures do depend upon visual clues from the ground below. They have been seen following the indentations and outcroppings of shore lines. No doubt they also find the meanderings of rivers valuable from a navigational point of view since these

The delicate wings and fantail of the Chinese bluepie are beautifully depicted in *Birds of Asia* by the famous British ornithologist John Gould.

make strong visual patterns against the green, brown or other colour of the land.

There are about 8,600 living species of bird and there is no area of the world which does not support some avian life. Some areas, however, are more abundantly populated than others. The North Pole has been visited by at least four species of bird, but only skuas visit the South Pole. The Antarctic continent supports 16 species of bird, mostly along its shores and on its islands but also many miles inland.

Other poorly endowed areas are oceans, deserts, or systems of islands (archipelagos). Few birds inhabit islands because remoteness has made them difficult for the

Flamingoes in Europe and Africa live in large flocks in shallow water. They possess a beak for filtering food from the water and can stand for a long time on one leg.

birds to reach from the larger land masses, and thus colonization has been restricted.

The number of species increases considerably in the temperate areas of the world – Great Britain and Ireland, for example, have about 450, Japan has 425. But the really big counts are reserved for the densely forested areas of the world like the Congo and the Amazon belt of South America. The country with the largest share of the world's bird life is Columbia, where a total of 1,700 bird species have been noted. In their splendid and comprehensive guide to ornithology, *The World of Birds,* James Fisher and Roger Tory Peterson call Columbia 'the heartland of ornithological variety on our planet'.

This distribution pattern has naturally required considerable special adaptations in the animals; it is instructive to look briefly at the subdivisions of the class from this point of view.

Some 600 living species are adapted for life on or in water. They include the beautiful and stately flamingoes. These birds stalk through the shallows of the warm African lakes repeatedly dipping their strongly curved bills into the water to feed. They suck the liquid into their beaks in a steady stream and eject it from the corners of their mouths. On its way it passes through a filter system which collects the food content. Similarly, the heron moves about in the shallows on its stilt-like legs, but is able to spear luckless fish and other small animals with rapier-like thrusts of its sharp beak. Webbed feet are another modification for life in water – to be seen commonly in the swans, geese and ducks.

Life without land

Most of the water birds of this type feed by poking their heads beneath the surface of the water; others bob right under or use their webbed feet for swimming beneath the surface. They may be carnivores

The great crested grebe flies poorly and is unhappy on land. If the female is disturbed on her nest she covers her eggs and then dives into the water.

or vegetarians; an interesting example of the former, and one which is specially adapted, is the oyster-catcher. Its bill is so constructed that is can be used like a powerful chisel to knock limpets and molluscs from the rocks.

Most of these birds confine themselves to fresh water but there are about 260 further species which have taken to life at sea. In some instances adaptation has

A circling vulture is the sign of impending doom. The king vulture from southern America is a scavenger feeding on carrion. Its wing-span can reach over six feet.

gone so far that they drink only sea water and spend many years without touching down on dry land at all. An example of this is the albatross, but the majority, such as puffins and gulls, live at least part of their lives ashore. The cormorant is one of these but this bird has taken to the water with such relish that is uses its wings like giant fins to swim about beneath the surface in search of food. In the East, incidentally, these birds are used to catch fish for fishermen. The bird is attached to a thin line so that it can be recovered once it has dived; and it is prevented from eating its catch by a ring placed around its neck which stops it swallowing.

The birds of prey are divided into two groups: the day hunting types, or *Falconiformes;* and those that operate at night, the *Owls*. These predators generally have very powerful talons, with which to grab their prey, and sharply hooked beaks with which to tear it up if it cannot be eaten whole.

The daylight hunters are broad winged and soar to great heights, dropping like stones to take their victim on the ground, or in some cases, if it is a smaller bird, while still in flight. The daylight group includes eagles, ospreys and falcons, but there are others less well endowed with talons – vultures, buzzards and the like – which feed on dead meat, carrion. The group varies greatly in the size of individual species. Apart from the large ones already mentioned it also includes the tiny falconet, which employs all the fierce skills of its bigger cousins to seize and kill insects.

The owls dispose of their victims in the same way as the daytime predators, that is with beak and claw. But obviously they have eyesight which is adapted to seeing in the dark and their hearing is extremely acute, allowing them to assess the range and direction of sounds.

The most brightly coloured of all the birds are to be found in the equatorial forests, and a great many of these belong to the fruit and seed eating group. They exist in great numbers because the forests

Long-tailed macaws from Central and South America, the largest and most colourful parrots, are good 'speakers'. Note the nut-cracking beaks.

provide a never-ending supply of their dietary requirements. Seeds and nuts often have hard shells so the nut-cracking species, which include some parrots, have a strong beak with sharp edges to do the job. Much more remarkable are the birds which live, at least in part, on honey. Among these are the humming-birds,

Dull brown in colour, the emu is the second largest living bird — but it cannot fly. It inhabits the Australian deserts. The male builds the nest, incubates and broods the chicks.

Reminiscent of a sentry dressed in the traditional uniform of some exotic state, the Californian tufted quail also keeps watch for enemies, its plume erect on the crown of its head.

which carry out their feeding while still in flight; they push their beaks into the flower as they hover above it and suck out the honey. The sword-billed hummer is found in the Andes and has a five-inch bill which is longer than the rest of its body.

A further division in the family of birds is provided by the *omnivores* — those that will eat vegetable matter, insects and even small mammals. In Britain, for example, the skylark eats both insects and the seeds of weeds, a relatively sedate diet; but the scarlet-rumped tanager from Central America is much more adventurous. Its intake ranges from bananas, through spiders, to mice and eggs.

Birds which feed on insects also occur in great numbers. Many of them hunt on the wing and are capable of great speed and manoeuvrability. Among these are the swifts and swallows, both of which fly enormous distances back and forth over the same area during a day's feeding. Most birds give the wasp a wide berth when it comes to eating, but not so the shrike, which has learnt how to pluck the sting out of the insect before it is eaten. This is, of course, a behavioural adaptation rather than a physical one. Also of interest here, are those birds like the South American ant pipit and tanager, which follow the ant armies and feed on both the ants themselves and the other insects which the marching column disturbs.

Birds for food

Next, there are the *ratites,* or flightless birds, of which there are 46 living species. The ostrich, emu and kiwi are included in this group.

The game birds are much hunted by Man as a source of meat. They include the pheasant, partridge, grouse and quail, and the turkey. Some of these are able to fly efficiently but others are less well equipped for aerial manoeuvring. Most of them are very quick on their feet and often use this method combined with short flights to evade danger.

Other birds used for food were long ago domesticated by Man. Most of these have been bred specifically to provide meat and are ill-adapted for life in the wild.

The ability to fly has enabled birds to conquer both land and ocean throughout the world. They have adapted in many different ways to meet the challenges of colonization, but there are many qualities which even now we do not understand.

Mammals — the lower orders

From duck-billed platypus to Man, the variety in size and structure of the mammals is astounding. Even the egg-laying monotremes and pouched marsupials have the ability to suckle their young.

HUMAN BEINGS, together with most of their domestic animals, belong to the class of animals known as *Mammalia,* or mammals. Members of the group range in size from the gigantic blue whale, the largest animal ever known on Earth, weighing up to 120 tons, to the tiny tree shrews, not much more than an inch long and weighing about three grams. Mammals are found all over the world – polar bears and arctic foxes have been found close to the North Pole, while whales, seals and dolphins are adapted to life in the cold oceans. Generally regarded as the most advanced animals, mammals are *craniates;* that is, they have skulls and backbones, but differ from the amphibians, reptiles and birds, by possessing a larger brain, hairy skin and an ability to suckle their young. It is this last ability that gives the mammals their name (Latin, *mamma,* breast). Like the birds, mammals have a four-chambered heart – venous blood flows into the right side of the heart through the right atrium and ventricle and to the lungs. The blood then returns from the lungs bearing oxygen and is pumped to the tissues by the left ventricle. Apart from three species (the platypus and two ant-eaters) the mammals produce their young alive. All are warm-blooded, like the birds but unlike the reptiles, and have elaborate internal mechanisms for maintaining body temperature within a very narrow range.

Learning how to live

A large brain means that a much greater proportion of behaviour is learned than is the case with other animal types. In birds, for example, the brain is smaller and less developed than the typical mammalian brain. This restricts the bird very largely to instinctive behaviour. The young mammal requires a considerable period of protected life while its parents and other members of the species teach it to cope with enemies, to find food, to seek a mate. The role of instinct decreases in the higher mammals, playing a relatively minor part in Man.

Control of body temperature means that the body processes take place in a constant environment. The kidneys and lungs ensure that the acidity of the tissues is maintained within a narrow range; the kidneys are also responsible for conserving the salt content of the body. This

Domestic cats have litters of five or six kittens. They are fed from eight abdominal teats and at birth are blind and completely dependent on their mother for food and protection.

temperature stability makes it possible for mammals to develop a more sophisticated biochemistry than is possible for lower animals.

As soon as they are born, the young are fed with milk manufactured inside the mother's body. This makes them to a large extent independent of the availability of food in the environment and they can rely on the mother to provide the essentials of life until they are able to feed themselves. This period of dependence is essential for animals that must be trained by their parents.

The four-chambered heart is an extremely efficient way of ensuring that the blood is washed completely free of carbon dioxide in the lungs, and that all the oxygen accumulated by the blood in passing through the lungs is made available in the tissues throughout the body.

More than 8,500 species of mammals are at present alive on the Earth. The history of mammals is unclear, but it is certain that they are descended from reptiles. One group of reptiles, the *Synapsida,* appeared on the Earth about 60 million years ago. They were heavy animals, some reaching the size of large dogs. These animals were abundant in the Permian period, about 30–40 million years ago. The bone structure of fossilized specimens shows that they were extremely similar to primitive mammals. It is difficult to be certain about the ancestry of the mammals because the only evidence of the early reptiles is in fossils. There is no way of knowing whether these animals suckled their young, had four-chambered hearts, were hairy or warm-blooded. However, the weight of evidence suggests that mamalian origins were among the Synapsida.

The difficulty of determining the early history of the mammals makes it difficult to classify modern mammals according to their evolutionary history, but zoologists are now generally agreed on the basic classification of the mammals. The living mammals can be divided into three sub-classes.

A New Forest pony suckles her young foal. The composition of milk varies from mammal to mammal - cow's milk is the most balanced.

The sub-class *Prototheria,* which is now largely extinct, has three surviving species. All three are *monotremes,* and are extremely primitive. All are confined to Australia and New Guinea and are very different from the rest of the mammals. Although they have milk-glands, hair and a relatively large brain, they lay eggs. The platypus, commonly but inaccurately referred to as 'duck-billed', is the best

The duck-billed platypus is an egg-laying mammal. It is about 20 inches long with webbed feet for swimming, a beak for grubbing for food but no pouch. The young are taken into a fold in the mother's skin where milk is discharged.

known representative of this sub-class. It is confined to Australia, and lives in pools and streams. Although the platypus has some ability to control its body temperature, this ability is not well developed; its temperature often falls by as much as 15 °C. Its close relatives, the spiny anteaters or echidnas, are found in New Guinea and Australia. They live on ants, lay eggs like the platypus, and have no teeth; the upper part of the body is covered with stiff spines, and the skeleton, like that of the platypus, is very different from other living mammals. All three monotremes have probably been able to survive because they have few natural predators and are highly specialized for their particular habitats. The platypus was at one time hunted almost to extinction because

of its fine fur, but it is now rigidly protected by the Australian government.

The sub-class *Allotheria* is now entirely extinct, but it is thought that this group of mammals evolved independently of the modern mammals. They existed for almost 70 million years, and fossil evidence shows that they were quite successful. Their ecological position may have been similar to modern rabbits.

Competition for space

The sub-class *Theria* contains three groups, one of which is totally extinct. The other two groups are the *marsupials* and the *placental* mammals. The latter contain the vast majority of mammalian species and are the most highly developed.

The marsupials are now confined largely to Australia and South America. They are basically similar to the placental mammals, but their young are born in a less developed condition and usually finish their development inside the mother's pouch. Some marsupials, however, do not possess a pouch; in others the development of the young before birth is very similar to reproduction in placental mammals.

The oldest fossil remains of marsupials have been found in Canada, but the animals were fairly common all over the world until about 15 million years ago. In the Old World they have been eliminated by competition with the more highly developed placental mammals; in Australia, however, marsupials had few competitors until the arrival of Europeans and they have developed a considerable variety of species. Although they have evolved separately from the mammals of Europe, Africa and Asia, the marsupials of Australia and New Guinea have in many cases developed species which have close counterparts in the Old World. This phenomenon, known as *convergence,* is a striking illustration of natural selection. Some

marsupials came to fill the equivalent position of mice, cats and moles, and many remarkable resemblances between these animals and their marsupial counterparts can be found.

Even in Australia, isolated by the sea from the rest of the world, the marsupials were not entirely left to themselves. Bats flew into the area, while rats and other rodents arrived, perhaps on pieces of drift-wood. These rats flourished and gave rise to a large number of native species. They probably eliminated some of the smaller marsupials by competing with them for food and living-space. Originally, there was a marsupial species similar to the wolf, but by the time Europeans arrived on the Australian mainland, the marsupial wolf was extinct, having been ousted entirely by the dog or dingo. In Tasmania, however, there were no dogs, and here the marsupial wolf still flourished

1 Like the platypus, the echidna or spiny ant-eater is an egg-laying monotreme. It has strong claws and a sensitive snout to search out termites and ants. When alarmed it can roll into a prickly ball or burrow into the ground. It lives in rocky areas of Australia and New Guinea.

2 The dingoes are believed to be the descendants of domestic dogs brought to Australia in prehistoric times. Today they prey on sheep and in this case a not-so-agile wallaby.

at the time of the European colonization. The Tasmanian aboriginals themselves are now extinct, but a few of the marsupial wolves have survived.

Agile kangaroos

The best known of the marsupials are the Australian kangaroos and wallabies. These animals are adapted for swift travel over land. They are vegetarians, equipped with powerful hind legs with which they move in a series of long jumps. Most have large pouches, or marsupiums, in which the females carry their young after they are born. At birth the baby kangaroos are extremely small, in some cases not much bigger than a man's thumb-nail, and are almost incapable of any independent activity. The kangaroo assists them into the

The kangaroo licks her fur to aid the migration of her newly born young to the pouch. After two months it can leave the pouch to find food but returns at all other times for a period of up to a year. Only one offspring is born each year.

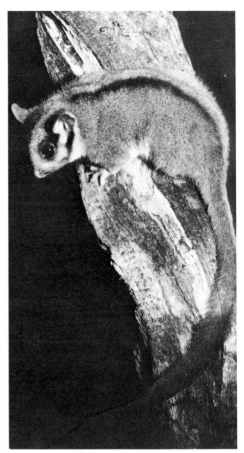

The Australian possums resemble dormice or squirrels and their habitat is the trees. This is the long-tailed Leadbeater possum.

pouch by licking her fur in a broad band from her vagina to the pouch. The tiny babies then 'swim' up the bank of wet fur. They remain in the pouch for as much as a year feeding on milk from nipples inside the pouch. Once the tiny foetuses become attached to the nipples they swell, thus ensuring that the offspring do not become accidentally displaced from the pouch.

The largest of the kangaroos is the

The 'badger' of Australia is the common wombat. It is a nocturnal, herbivorous animal which digs burrows up to 100 feet long. The female raises one baby at a time.

great grey kangaroo which reaches a height of six feet; the smallest is the rat-kangaroo, about the size of a hare.

In the same general zoological grouping as the kangaroos are the Australian possums, the flying phalangers, the wombats and the koala bear. The large family of possums includes a number of tree-dwelling species. Some, such as the sugar glider, have long feathery tails and webs linking their front and hind limbs. They are able to glide from tree to tree. The possums were once extensively hunted for their beautiful fur, but are now protected and have become much more common.

The wombats, heavy and rather sluggish vegetarians, live in underground burrows and emerge at night to feed.

The koala bear, the original 'teddy bear', is a sluggish tree-living animal, and not a true bear at all. It will only eat the leaves of a few species of eucalyptus tree. It never drinks, but absorbs sufficient water from the leaves on which it feeds. It does, however, eat earth, presumably in order to obtain essential minerals. The pouch of the female koala is downward facing and the young are born only 34 days after conception. They have to make their way without assistance to the pouch; after eight months the baby emerges as an eight-inch youngster. For some time afterwards, the baby koala is carried around by the mother, on her back or in her arms.

'Playing possum'

The Tasmanian wolf is representative of a large group of flesh-eating marsupials found in Australia. Some of these are similar to the cat family, and include the so-called tiger-cat. These animals are easily able to hold their own against 'imported' cats and dogs, but are hunted down where possible because of their ferocious raids on poultry farms.

The South American marsupials are somewhat removed from the Australian

49

The dark-grey fur seals are protected from the cold by a thick layer of fat. These aquatic mammals look after their young for four months — but many do not survive their first year.

1 The common opossum is the United States' only marsupial. It has two litters of nine each year; the young stay in the pouch for three months, then travel on their mother's back.

2 The offspring of an opossum still in the pouch. They crawl there a short time after conception when only half an inch long and feed from internal nipples.

3 On the alert for danger – a lioness and her cubs. Two or three cubs are born at a time and are weaned at six months. After five months they accompany the male on his hunting trips.

species although both groups probably arise from a common stock. The American species have undergone a process of selection in competition with placental mammals. At one time there were marsupial 'bears', carnivores rather like the modern grizzly bear, but these were ousted by similar placental species. There are also fossil remains of a marsupial superficially similar to the sabre-toothed tiger.

The best known of the American marsupials are the opossums. They occur all over the American continent, live in trees and feed on insects. Some of the species of opossums in America have the remarkable habit, not shared with their Australian counterparts, of shamming dead when threatened. This is the origin of the expression 'playing possum'.

The survival of these animals that have not achieved the level of development of the true mammals is certainly due to geographical isolation in the case of Australia and New Guinea. In all probability South American varieties have not been extinguished because of the late arrival of true, or placental, mammals to the continent.

Mammals —
man's closest relatives

At the summit of the animal kingdom are the true mammals. Their young are born in a more advanced condition than any other group — a common feature which classifies a wide variety of creatures.

THE PLACENTAL or true mammals form the largest and best developed group of Mammalia. They are generally regarded as the highest point of the evolutionary tree. Man is a placental mammal, as are the great apes, the whales, all the mammals of the Old World and most of the mammals found in the Americas.

This group of animals are called 'placental' because their young are carried in the mother's womb and born alive, the blood circulation of the unborn young being linked with the blood-stream of the mother by a complex of membranes. These membranes, bathed on one side by the mother's blood and on the other by the blood of the offspring, transmit substances required for the growth of the offspring, and carry away its waste products. This system of membranes is called a *placenta*.

Ancestors of the mammal

The early placental mammals appear from fossils to have been rather similar in appearance to today's shrews. Zoologists believe that these early mammals lived in trees, as indeed some of the shrews do today. For reasons that are not fully understood, the dinosaurs and large reptiles disappeared rather rapidly from the Earth, while at the same time there was a rapid development of the mammals. They spread quickly across the surface of the Earth and diversified into a large number of species and types.

The early mammals probably lived largely on insects, as moles and ant-eaters do today. Modern mammals feed on an astonishing variety of foods: *carnivores* like dogs and cats eat the flesh of other animals; *herbivores* like cows and elephants eat grass, tree leaves, and other vegetable matter; *rodents* – rats, mice, squirrels – also live mainly on vegetable matter but obtain their food by gnawing rather than by chewing; *omnivores* such as Man will eat almost any food available. Of the aquatic mammals, seals eat fish, sea cows eat only sea plants, others like the whalebone whales, live on plankton.

Adapting to their needs carnivores have become well equipped for catching their

The 'tiger' of South and Central America, the jaguar, hides during the day and hunts at night. This ferocious big cat can be a man-eater.

prey with sharp claws, strong teeth for tearing flesh, and an ability to run faster than their prey. Most have developed good eyesight or sense of smell for tracking down the animals they wish to eat. Herbivores, on the other hand, often require special internal adaptations in order to digest their food. The four stomachs of the cow, for example, make it possible for it to digest cellulose, the main constituent of grass. Cellulose is normally indigestible for carnivores because they lack the necessary enzymes to break down this substance in their stomachs.

Escaping the killer

Other vegetarian animals are well adapted for flight from their predators. The gazelle, which is preyed upon by the lion and hyena, has long legs and quick reflexes to enable it to escape. Other herbivores, like the elephant and the hippopotamus, are large and powerful enough to deter any meat-eater.

The carnivores alive today are generally separated by zoologists into two main groups. One group contains cats and similar animals; the other, dogs, bears, weasels, and animals like them. The cats are extremely pure examples of carnivores; their teeth are adapted solely for tearing flesh and they have no equipment for chewing their food. This restricts them entirely to flesh, because flesh can be swallowed whole and digested, whereas vegetable matter cannot. Cats tend to be individualists, unlike dogs, and do not hunt in packs. They cannot run for long distances and catch their prey by stalking and springing on them. Lions and tigers are the best known big cats. Lions and tigers have been successfully interbred, notably at the Paris Zoo, producing 'tigons'

Common red foxes generally live in burrows, or earths. During daylight they lie low and at night hunt birds, rabbits, mice and moles.

with some characteristics of both. These artificial hybrids are, however, sterile. A mammal closely related to the cat is the hyena. This animal was once thought to be a scavenging species; but detailed studies of hyenas in African national parks have shown that these animals do not generally scavenge. They catch living prey, such as antelopes, by hunting them down.

Unlike the cats, the dogs, foxes and wolves have retained some chewing ability and have some molar teeth. They tend to hunt in packs and can run for long distances. Some of the wild forms found today, such as the Australian dingo, are probably descended from domesticated dogs that have reverted to the wild state. The bears are closely related to the dogs and are also capable of eating a mixed diet. The only purely carnivorous bear is the polar bear, which has little choice in an area where fish is the the only available food.

The marine carnivores are probably descended from land carnivores of the dog type. They include the seals, which live almost exclusively on fish, and the walrus, which uses its huge canine teeth or tusks to open the shellfish on which it feeds. In these animals, the legs have become adapted for swimming.

The most adaptable and successful of all mammals are not the carnivores, but the rodents, the gnawing mammals. Best known are the rats and mice, but the family also includes squirrels, porcupines, rabbits, chipmunks and guinea pigs. Rodents are found in every part of the world, and some live in close connection with Man. The rodents are almost entirely vegetarian. Characteristic of the group are prominent incisors or front teeth which grow at a rapid rate. If the animal for some reason ceases to gnaw for a period, the growth of the incisors tends to force the jaws apart and the animal may die.

Many rodents dig burrows in the ground; some, such as beavers and musk-rats, are

1 Barely two inches long, the harvest mouse is an agile little mammal which uses its prehensile tail to perform fantastic balancing feats.

2 Shrews are very small insectivores and have a three-hour cycle of sleeping and feeding. The common shrew, like other species, lives alone.

3 Wild rabbits are smaller than hares; they live in large groups in complex warrens.

4 The incisors of a beaver are used for gnawing; flaps of skin prevent wood chips from entering the mouth. Teeth grow quickly to compensate for wear.

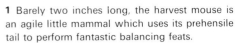

partially water-dwelling. Others live in the trees, and while there are no flying rodents, some of the squirrels can glide long distances using webs of skin between their limbs and trunks. The extraordinary ability of rodents to adapt is shown by the rats. These animals were the only placental mammals to enter Australia unaided by Man, and have developed a number of forms well suited to their new environment.

The guinea pig and the porcupine are representatives of a group of rodents which penetrated into South America when North America was a separate land mass. They include the largest living rodent, the capybara, or water-pig, which grows as large as a domestic pig. Hares and rabbits although successful are unlike the other rodents because they have four upper incisor teeth.

The *ungulates* contain the mammals which are most useful to Man. This is a large and diverse group of animals developed from different evolutionary lines. Almost all the large herbivorous mammals are ungulates: cows, horses, sheep, elephants, camels and other animals domesticated by human beings. The ungulates are not a single group and are classified together more for reasons of convenience than because all the animals are closely related. All, however, have certain simi-

The American black bear is a quiet animal, a good climber and is partial to fish. Protected in national parks, it is elsewhere hunted as a pest.

larities, which are chiefly related to their method of feeding.

The obvious similarities are in the teeth of the ungulates. Their food is generally grass or vegetable matter which must be thoroughly chewed before it enters the animal's stomach. This has given rise to molar (grinding) teeth of large size and surface area.

Swift-toed horse

The ungulates are usually highly mobile, either to escape from carnivores, or because they must traverse a wide area to find food. Many African species, the antelopes for example, may travel thousands of miles in the course of the year and migrate from winter feeding grounds to summer pastures. The horse is a typical example of an ungulate well adapted for rapid motion, but a fallen horse finds it difficult to rise. The horse's legs are typical of many fast-running ungulates; the first part of the limb is relatively short making it possible for the second, longer

At night, the Brazilian tapir feeds on leaves and fruit with the aid of its prehensile muzzle. It is an odd-toed ungulate having four toes on each front foot and three on the back.

segment of the limb to move rapidly, giving a powerful muscular thrust to the drive of the leg. The third section of the horse's leg corresponds with the hand or foot in Man. In animals like the horse, the bones of the foot are lengthened and the animal really runs on its toes. The horny hooves are adaptations of the claws found in the early primitive mammals. Claws, where they do persist among the ungulates, are generally blunted and have the function of protecting the feet.

The horn is typical of the defence mechanisms developed by the ungulates against hostile predators. Another defence, developed, for example, in the pig family and by the elephants, is the elongated tusk. In reality this is a canine tooth, valuable as a weapon of defence or attack.

The ungulates are divided into odd-toed and even-toed; typical of the former is the horse. It runs on the middle toe, the other toes being reduced in size and no longer used in running. Three-toed horses are found in fossil remains, but today this type of foot is found only in the rhinoceros. The only truly wild horse now living is the rare Prezwalsky horse of the Mongolian steppes, although there are other species, like the native ponies of the British Isles, which have reverted to a semi-wild existence. The other major representative of the odd-toed ungulates is the tapir of Malaya. This animal retains a very primitive foot formation; four toes on the front feet and three on the hind feet.

The even-toed ungulates are represented by a much wider variety of living species than their odd-toed relatives. They include the diverse cattle family, with such wild species as the buffalo of North America. To these are related the deer and camel families and the giraffes. They form a large group of *ruminants*, pure vegetar-

The 'white' rhinoceros is, in fact, a muddy grey in colour. Despite its four tons weight and menacing horns it is quite placid unless provoked.

Deer have less developed hooves than other ruminants and are still four-toed. This red deer stag is 'in velvet' - growing new antlers.

ians with a complex arrangement of stomachs for digesting grass and leaves. After being taken into the mouth, the food travels into one of the stomachs where it it is partly digested and then regurgitated to the mouth to be chewed over again. The camels and their relatives the llamas

are descended from North American species and there were native camels in the United States until quite recent times. The giraffe is another ruminant; the formation of its long neck does not involve the addition of any bones to the animal's skeleton. Almost all mammals have seven neck bones; the giraffe is no exception, its bones are just extremely long.

The elephants belong to another group of ungulates, or more properly sub-ungulates. Their group, the proboscidians, was originally widespread in the Northern hemisphere where mammoths were once common animals. Well-preserved mammoths are found regularly in the frozen bogs of Siberia; it has been reported that the flesh is often still edible. The proboscidians also include another bizarre species of mammal – the dugong or sea-cow. These belong to the *sirenians,* so-called because the appearance of these vegetarian beasts off tropical coasts probably gave rise to legends about mermaids. Despite their specialized habits, the sea-cows are closely related to the elephants.

Whales are the largest group of aquatic mammals and the best adapted to life in water. They are able to withstand the major pressure changes involved in deep diving and are insulated against cold by a thick layer of fatty blubber. They are also streamlined for fast and efficient movement. Some of the whales appear to have a very high degree of intelligence, and species like the dolphin are friendly towards human beings, easily learn complicated 'tricks', and appear to have a well-developed method of communication.

Primates – born in the trees

The most intelligent of the land mammals are the primates, the group to which Man himself belongs. The lemurs, monkeys and apes are derived from tropical tree-living forms, and indeed most of the present-day primates still live in trees. They are omnivorous, eating both flesh and vegetation,

The monkey's limbs allow it to scamper across the ground; its human-like hands allow it to climb trees; its long, limber tail gives it the ability to swing from branch to branch.

Lemurs, or 'half-apes', are the most primitive primates. The ring-tailed lemur of Madagascar prefers thinly wooded country to thick forest areas.

although many species rely more or less exclusively on vegetable food. Their mode of life in the trees has given rise to a well-developed sense of sight, and the necessity for quick reactions and muscular co-ordination in order to swing around the forest has given rise to a highly developed brain. The lemurs are generally considered to be the most primitive of the primates. The monkeys are divided into Old and New World types. The New World monkeys are less advanced from an evolutionary point of view than the Old World monkeys and are chiefly distinguished from them by their flattened snouts. The Old World monkeys include the great apes and Man. The most developed of these are the gibbon, orang-utans, chimpanzees and gorillas. Of these, the last two have relatively large brains and a fair degree of intelligence. Both are large animals, and both have begun to move out of the trees and on to the ground.

But the highest form of primate, *Homo sapiens*, is another story altogether....

Index